STATE OF THE

WARATAH

THE FLORAL EMBLEM OF NEW SOUTH WALES IN LEGEND, ART & INDUSTRY

An illustrated souvenir

Editor Rosie Nice

PUBLISHED BY
THE ROYAL BOTANIC GARDENS SYDNEY
ON THE OCCASION OF THE EXHIBITION
STATE OF THE WARATAH
AN OFFICIAL
OLYMPIC ARTS FESTIVAL EVENT
SYDNEY 2000

Above
Wood panel c1910
School of L.J. Harvey
40 x 14cm
KEITH FREE COLLECTION

Page 3
An Australian study
Postcard 1906
ROBERT HUTCHINSON COLLECTION

THIS MAGNIFICENT AUSTRALIAN

Back in the fifties and sixties I recall being embarrassed by the sight of dinky-di Aussie images. The Early Kooka stove was seen as a badge of backward-looking suburbia. 'The Bridge' on every packet of Sydney Flour seemed unutterably parochial. As to the waratah, embossed on a thousand Wunderlich pressed metal ceilings, carved into chests and sideboards, fretted out of the slats at the back of dining chairs, scorched into endless pokerwork breadboards and turned wooden vases, it seemed an unspeakably stiff and unyielding flower. But tastes change. We become wiser. Now I see the waratah as one of the most beautiful creations of nature I can imagine, and can understand why it captured the imagination of our forebears, inspiring designers and craftspersons to appropriate its globular form to decorate everything from a tile to a cupola.

As a garden plant it was especially prized. The story goes, and I believe it, that the first waratahs successfully grown in a Sydney garden were those planted at Bronte House by Hugh Beattie, Dame Mary Gilmore's grandfather. Beattie was the bailiff, manager we would call him today, there when the house was occupied by Robert and Georgiana Lowe who built it in 1845. Mrs Lowe was a keen gardener and it is said that she brought not only the plants from the bush but a dray-load of native soil to ensure that they flourished. I have planted one here in the same spirit as she did, to celebrate this magnificent Australian. Long may it flourish and bloom.

Leo Schofield
Bronte House, Sydney

IN THE BEGINNING

Long before European settlement, Aboriginal people saw the bright red flower which shot up from the forest floor and called it 'Warada', red flowering tree. Scientists tell us that the waratah had its beginnings over 70 million years ago in Gondwana.

John White, surgeon on the First Fleet, sent a waratah specimen to London in 1791 and the first botanical drawing was published in 1793. The aptly named Telopea speciosissima, meaning most handsome seen from afar, is now the State floral emblem of New South Wales.

A movement to encourage the use of Australian flora by Australian designers was led by Lucien Henry, who arrived in Australia in the 1870s. His extraordinarily imaginative designs included waratah columns for Sydney's Hyde Park.

In 1906 curator Richard Baker initiated an exhibition 'Australian Flora Applied to Art'. Referred to as the 'Commander in Chief of the Waratah Forces', Baker, in 1911, sent waratahs to the Melbourne Town Hall where the National Flower was being debated and the waratah lost to the wattle.

Artist Ellis Rowan captured the beauty of our native flora in 'Waratahs New South Wales' painted in 1882. In 1916 Eirene Mort designed a waratah logo for the Arts and Crafts Society of New South Wales. Artists Margaret Preston and Bruce Goold have favoured the waratah in their prints and designers Linda Jackson and Jenny Kee revived its popularity in 'Art Clothing'.

The waratah features in the logo of many New South Wales-based organisations. It has been used in the designs for pressed tin, plaster ceilings and wrought iron balustrades, as motifs in leadlight and etched glass, as a feature on public buildings and as ornamentation in private homes.

'Preservation by cultivation' is the catch-cry of waratah lovers everywhere. The new cultivars are more spectacular and easier to grow. Renewed interest in the waratah is highlighted by the revival of the Waratah Festival, the inauguration of a White Waratah Auction at the Sydney Flower Markets and the advent of waratah farms.

Once again, in the new millennium the waratah is being used in design and decoration. The Sydney Olympics has thrust our magnificent waratah on the world stage; an international star that surely will never wane.

Page 5
Lucien Henry (1850–1896)
Hippocampus & waratah fountain 1888–1891
Illustrated bookplate
Watercolour
One of a series for a proposed 'Australian Decorative Arts' book
POWERHOUSE MUSEUM, SYDNEY

ABORIGINAL DREAMTIME STORIES

Aboriginal history is an oral history.

Tales are told and retold so that today we have many stories. They come from the same source. The details may differ but the similarities confirm our suspicions that these are not myths but the truth.

To the Dharawal people, whose lands run from Botany Bay south, the Dharug people who live from the west of Sydney to the mountains and the Gundungurra people from the south west of Sydney, the waratah has special significance. Throughout history the waratah has been part of their lives whether drinking its sweet sustaining nectar or using the leaves as a protection against fire.

Waratahs in the wild 1999
Photo Jaime Plaza
ROYAL BOTANIC GARDENS SYDNEY

Long ago there lived a beautiful Aboriginal maiden named Krubi. She made herself a cloak of the red skin of the wallaby, and ornamented it with the even redder crests of the Gang-Gang Cockatoo. It was said to be the only cloak of its kind in the world.

Krubi knew a young man who was far enough removed in blood to be encouraged in his love for her. She always chose a cleft between two great weathered sandstone rocks on the top of a ridge to await the return of her man from the hunt. The red figure was the first object to strike the eyes of the returning hunters and the red cloak was the only object looked for by the young man. One night Krubi's heart was saddened. She learned that men from another tribe had been seen in the valley and there was to be a great corroboree, during which her man would be taught what to do in war. The day of the battle, Krubi stood in the sandstone cleft and watched for the return of the warriors. She heard the yells of war. In the afternoon she saw the scattered tribe of men return, but no lithe young figure stepped out from the others to greet her.

Krubi stayed for seven days waiting, and her tears formed a little rivulet. Then she went back to the camp but it was deserted and the ashes were cold. Krubi returned to the ridge, and willed herself to die. She passed into the little cleft of weathered sandstone, and up came the most beautiful plant. The stalk was firm and straight, without a blemish, just like the man Krubi died for. The leaves were serrated and had points like a spear. And the flower was red, redder and more glowing than any other. It was the first waratah.

EXOTICA FROM THE ANTIPODES

From the beginning of colonisation, Europeans admired the beauty of the waratah. According to James Edward Smith, noted botanist: 'The most magnificent plant which the prolific soil of New Holland affords is, by common consent both of Europeans and Natives, the Waratah.' Artists from the first responded to its theatricality and novelty. The colony's first governor, Arthur Phillip (1738–1814), also commissioned an early drawing of the waratah. He told botanist Sir Joseph Banks, who was in England, that on 3 December 1791 his collector had packed for him several of the plants in tubs, and that he also was sending 'a drawing of the War-ret-tah'.

It is not known what happened to this drawing, but one 'coloured drawing made from the wild plant', possibly from the specimen sent back to England by Surgeon-General John White, was used by artist James Sowerby for Plate Seven — *Embothrium speciosissimum* — of Smith's *A Specimen of the Botany of New Holland*. This was a book intended for European naturalists fascinated with the worlds unfolding in recently colonised corners of the globe.

John Hunter, First Fleet ship officer and later governor of New South Wales, dismissed his own abilities as a naturalist and artist, describing himself as 'wholly unqualified to describe the colony's flora and fauna'. Hunter's drawing of the waratah bears out this analysis. Perhaps the first European drawing of the flower, it is more emblematic of its splendour than scientific. Significantly Hunter labels the drawing 'wa-ra-ta' — apparently the earliest adoption of its Aboriginal name. *Richard Neville*

Left
John Hunter (1737–1821)
Wa-ra-ta 1788–1790
Watercolour 22.6 x 18.3cm
REX NAN KIVELL COLLECTION
NATIONAL LIBRARY OF AUSTRALIA,
CANBERRA

Page 9
James Sowerby (1757–1822)
Embothrium speciosissimum
1793
Hand coloured engraving
23.5 x 16cm
From James Edward Smith's
*A Specimen of the Botany of
New Holland*
ROYAL BOTANIC GARDENS SYDNEY

Embothrium speciosissimum.

COLONIAL ART

That the waratah maintained its early pre-eminence over colonial flora, is suggested by this watercolour by P.H.F. Phelps. Phelps, who owned a property at Broulee, on the New South Wales far south coast in the 1840s, included this watercolour in a sketchbook he compiled of local animals and some Aboriginal subjects.

Phelps' charming drawings would not have satisfied a naturalist; rather they suggest the pleasure of an artistically competent enthusiast. Similarly, denoting the waratah a 'wild tulip of New Holland' suggests a populist rather than educated interest.

While Smith and Sowerby saw the waratah as a curiosity of a new land, colonists were proud of its dramatic beauty. Sir James Martin, in his Australian sketchbook of 1838, recalled youthful weekends sallying 'forth to a romantic part of the country' collecting specimens including immense bundles of 'warrataws, or colonial tulips'. For him it was a flower which defined an Australian experience.

Similarly Louisa Anne Meredith, herself an artist, was enchanted in 1840 to encounter a stand of waratahs at Mount Victoria, in the Blue Mountains. In her *Notes and Sketches of New South Wales* (1844), she described them as 'a sisterhood of queens — a group of eight or 10 splendid waratahs, straight as arrows — tall, stately, regal flowers, that with their rich and glowing hue — seemed like the magic jewels we read of in fairytales'. *Richard Neville*

Left
G.C. Fenton
Watercolour
Possibly 1860s
12.8 x 17.5cm
NATIONAL LIBRARY OF AUSTRALIA, CANBERRA R438

Page 11
P.H.F. Phelps
Warrataù or wild tulip of New Holland 1840–49
Watercolour & ink
25.6 x 17.5cm
IMAGE LIBRARY, STATE LIBRARY OF NEW SOUTH WALES ZDL PX 58, F.49

Warratàu, or wild tulip
of New Holland

FLORAL ILLUSTRATION

Sisters Helena and Harriet Scott were Sydney's leading natural history illustrators and artists in the later part of the 19th century. The daughters of an amateur naturalist and impecunious gentleman, their lives revolved around the dilemma of being gentlefolk, women and poor. Poverty threatened their social status as much as their lifestyles. Yet they were proud of their skills, and continually pressed their reluctant father, and Sydney's mainly masculine scientific community, for recognition. Indeed Helena told a friend: 'Oh! You cannot think how thankful I am that my dear father allows me to place my names on the drawings. It makes me feel twice as much pleasure while painting them.'

Both Helena and Harriet were employed in 1880 by Sydney booksellers Turner and Henderson to produce — influenced by the awakening interest in Australia as subject matter for Australian artists and writers — The Australian Floral Series of 24 Christmas and New Year cards depicting local wild flowers. *The Sydney Morning Herald* of 9 December 1880 approvingly noted that with these cards, which were also issued as Our Native Flowers, colonists were 'beginning to provide locally for their own requirements in art', rather than simply importing English cards and designs. The paper also commended the accurate drawing and beautiful colouring of the Scotts' cards. An original pencil preparatory drawing for the waratah card is in the Mitchell Library.

The Australian Floral Series appear to be the first locally published, Australian-designed Christmas cards. Most of the flowers depicted — such as the Christmas bush and the waratah — were readily found in the Sydney region: only a few iconic flowers, like Sturt's Desert Pea, were from other parts of the country. *Richard Neville*

Left
Harriet and Helena Scott
Waratah and Native Currant
Australian Floral Series c1890
Chromolithograph based on a watercolour
IMAGE LIBRARY, STATE LIBRARY OF NEW SOUTH WALES 581.991/F

Page 13
Helena Scott (1832–1910)
Telopea speciosissima —
Waratah 1879
Pencil drawing
IMAGE LIBRARY, STATE LIBRARY OF NEW SOUTH WALES PXA 1710 F.120

120
Telopea
Speciosissima
"Waratah"
Sydney Dec'r 1879

BOTANICAL DRAWING

It was the passion of Joseph Maiden, government botanist and director of the Royal Botanic Gardens, to promote the study and knowledge of Australian plants — in part to assist their commercial exploitation — and his many publications were devoted to this end. 'New South Wales is crying out for illustrations of her indigenous flora' he declared in his *Illustrations of New South Wales Plants* (1907). Maiden employed a number of artists to illustrate his works: Edward Minchen, Henry Baron, his own daughter Mary, and Margaret Flockton.

Maiden was interested in accurately describing Australian plants and documenting their history. His eight-volume *The Forest Flora of New South Wales* (1904–1925) was both an anecdotal history of the plants and a botanical text. The first six volumes were illustrated by Flockton's line drawings. Flockton's precise plates, which include dissection drawings, functioned as diagnostic images rather than works of art, and were designed to assist botanical comparison and analysis rather than celebrating a flower's particular beauty. This is obviously not the case with her earlier charming watercolours.

Flockton, who was born in England in 1861, trained in London as a technical artist at the South Kensington Art School. She worked for a Sydney lithographic firm after she arrived in Sydney in 1880. In 1901 she commenced work as a botanical artist at the Royal Botanic Garden's Herbarium at the rate of two shillings an hour. *Richard Neville*

Left
Margaret Flockton (1861–1953)
The Waratah c1916
Pencil drawing
24.7 x 20.5cm
ROYAL BOTANIC GARDENS SYDNEY

Page 15
Margaret Flockton
Waratahs 1899
Watercolour 59 x 38cm
MRS LYLIE KEMPE COLLECTION

A MERE FLOWER PAINTER

When the judges of the paintings submitted to Melbourne's Centennial Exhibition in 1888 first delivered their verdict, they chose to award the First Order of Merit, not to well-known Australian artists Tom Roberts, Arthur Streeton or Frederick McCubbin, but to Mrs Ellis Rowan for her flower studies which hung in the Victorian, Queensland and New Zealand courts. The group of judges from 15 nations recorded that none of the other works were of sufficient importance to call for special comment, but that Rowan's masterly delineations of the Australian and New Zealand wildflowers have excited general admiration and have secured for her the 'highest honour'. Members of the Victorian Artists' Society were outraged and insulted, responding with stinging criticism about the unfairness of the award. Even today, Rowan's work is largely unrecognised in art gallery collections, her work having been preserved by museums and libraries.

For over 40 years, Marian Ellis Rowan (née Ryan) (1847–1922) travelled throughout Australia, New Zealand, New Guinea and America recording in painstaking detail the rich botanical diversity of native species. Her paintings, so genteel, so at home on drawing room walls, belied the hardships she endured. Like her English counterpart, Marianne North, she journeyed into remote areas in appalling conditions to capture a specimen in its natural habitat. Her paintings were revered for their accuracy by the scientific fraternity, including noted German botanist, Ferdinand von Mueller, and American botanist, Alice Lounsbery. Her studies of Australian flora, in particular the waratah, were copied by ceramic artists at the Worcester Royal Porcelain Company to decorate bone china for Prouds Ltd and Flavelle Brothers, for sale to the Australian market. *Margaret Betteridge*

Top
Worcester Royal Porcelain Co.
Great Britain, established 1862
Plate with waratah motif from
Mrs Ellis Rowan Series 1912
Porcelain, painted enamel,
slip cast
NATIONAL GALLERY OF AUSTRALIA,
CANBERRA

Bottom
**Ellis Rowan with Mary Moule
and Eric (Puck) Rowan on pony**
Photograph 1887
NATIONAL LIBRARY OF AUSTRALIA,
CANBERRA
BY KIND PERMISSION MRS VELLACOT

Page 17
Ellis Rowan (1847–1922)
Waratahs, New South Wales
1882
Watercolour, opaque white
highlights
65.7 x 39.2cm
ART GALLERY OF NEW SOUTH WALES
GIFT OF MR AND MRS F. STORCH 1992
PHOTO: SUZIE IRELAND FOR AGNSW

A NATIONAL STYLE

Lucien Henry (1850–1896), an expatriate Frenchman trained in architecture, formerly exiled political prisoner, is credited with the first serious attempt in Australia to create a national art form. Henry arrived in Sydney in 1879 as preparations were in full swing for the International Exhibition. Nationhood was not far away and the time seemed right.

Henry was passionate about the cause, strongly advocating his case in an article in *Australian Art* in 1888. For him, the idea that Australians still clung to the old world for architectural and decorative inspiration was an anathema. He was entranced by the possibilities that he saw in our indigenous species of plants and animals and argued that the waratah could become for Australia what the lotus and the palm had for the Classical world.

Henry not only wrote about the idea. He taught art to technical students, encouraging them all the while to draw from nature, ours not others; and he exhibited his own artworks. He even created a legend for the waratah, to fabricate a dreamtime past that could be viewed with reverence, respect and awe. Some of his works were highly charged with sentimentality and allegory, all subtle weapons in his campaign for a national style.

It is interesting to note that the Garden Palace, which was built for the International Exhibition, covered an area of the Royal Botanic Gardens Sydney which stretched from The Conservatorium to what is now the Cahill Expressway. Sadly it was destroyed by fire in September 1882. *Margaret Betteridge*

From far left
Lucien Henry (1850–1896)
Australian Legend: The Waratah 1891
IMAGE LIBRARY, STATE LIBRARY OF NEW SOUTH WALES Q823.8/H522.2/1

School of Lucien Henry
Plaster bust with terracotta finish c1893
KEITH FREE COLLECTION

Page 19
Lucien Henry
Waratah 1887
Oil on wood
51 x 35cm
ART GALLERY OF NEW SOUTH WALES
GIFT OF MARCEL AUROUSSEAU, 1983
PHOTO: RAY WOODBURY FOR AGNSW

DELIGHTFULLY BIZARRE

To show just what he meant by a 'national style' and how it could be translated into art, architecture and the decorative arts, Lucien Henry embarked on an ambitious project. Rather than adapting European form, he set out to inspire public thought about his concept, using wild flights of fancy to provoke reaction. He took everyday objects and transformed them into vehicles to support his argument. Waratahs became decanters, light globes and wallpaper designs. One of his more enchanting ideas was a Lovers' Walk for Hyde Park, concluding at a red-roofed rotunda shaped as a waratah flower.

His plan was to publish the book himself, and he secured a number of donations and subscriptions from public and private individuals. Entitled *Australian Decorative Arts*, it was to include 100 illustrations and instructions on how to execute his ideas along with scientific data and a discussion on the industrial arts. A shortage of funds prompted him to travel overseas in the hope of attracting investment in the idea in Europe and England.

Sadly Henry's vision was never realised due in part to the economic downturn of the 1890s which scuttled his hopes of financial support, but ultimately by his untimely death in France in 1896. Henry's illustrations were presented to the the Technological Museum, now the Powerhouse Museum, by his son in 1911 and they have continued to inspire generations of waratah enthusiasts. *Margaret Betteridge*

From top left
Lucien Henry Illustrated
bookplates:

Design for Electrolier
Watercolour 1888–1891

Design for majolica tile panel
Watercolour 1888–1891

**Design for wrought iron and
enamelled bronze gate**
Watercolour 1888–1891

Page 21
Design for wallpaper
Watercolour 1888–1891

POWERHOUSE MUSEUM, SYDNEY

WARATAH WOODBLOCKS

Margaret Preston (1875–1963), sometimes known as 'the wild woman of Australian art', is best known for her instantly recognisable woodcut prints, which were both modern and popular. She contributed much to Sydney's artistic style from the 1920s to the late 1950s.

A strong-handed vitality was her trademark. Block prints could be cut from the very wood of living trunks, frottaged from bushfire blackened stems. Her insistence on hand printing blocks by rubbing the back of the handmade paper, often with a wooden spoon, added a grainy texture to her defining blacks. The water-based ink provided the skeletal lace key-line for the vivid gouache hand colouring, fiery red flowers and sap-green leaves.

In 'Waratahs' (1925), the stylised symmetry of the arrangement has the decorative quality of Australian Art Nouveau, and the high relief of German woodcuts, William Morris wallpaper and the eccentric English Omega Workshop screens. 'Illawarra Lilies and Waratahs' is one of her largest prints, made up of twelve end-grain pieces braced together (blocks suitable for wood engraving tend to be very small). The working process was quite different. In this instance, an oil sketch on wood has been cut to print the 'painting', whereas the earlier image was designed as a powerful graphic.

In the mid 1940s Preston almost ceased producing woodcuts, 'quite trying on the hands', though continued to make fascinating prints in many media — 'masonite' cuts, stencil and monotype — for the rest of her life. *Bruce Goold*

Left
Margaret Preston (1875–1963)
Illawarra Lilies and Waratahs
1929
Hand-coloured woodblock
59.6 x 44.4cm
NATIONAL GALLERY OF AUSTRALIA
CANBERRA

Page 23
Margaret Preston
Waratahs 1925
Hand-coloured woodblock
42.6 x 30cm
NATIONAL GALLERY OF AUSTRALIA
CANBERRA

MODERN OR DECORATIVE ?

Grace Cossington Smith was a bravely modern experimenter, whilst Adrian Feint embraced European decorative styles.

Cossington Smith (1892–1984) had a great affinity with her surroundings, mainly painting subjects from her home, her garden and nearby cityscapes, most famously Sydney Harbour Bridge. Her waratah appears to be almost sculpted: patches of paint layered to build up a domed efflorescence of dancing colour. Australian author Drusilla Modjeska vividly describes Cossington Smith's floral paintings in her book *Stravinsky's Lunch*:

> She has taken us to them as they grow and has filled the frame with them so that we look into petal and stamen as if we could reach inside and touch the essence of flower-ness.

The waratah painting is one of a pair; the other is a flannel flower. Both were sold as a 'thank you' to an admirer in 1973 at the original 1930s price, still marked on the back labels, of two guineas each!

Adrian Feint (1894–1971) was an artist, decorator, illustrator and designer with a penchant for the exotic. Fashionably stylish, his woodcuts are recognisable for their balanced yet dramatic areas of black and white, textured and patterned in a highly ornamental manner. Painted cornucopia of flowers and his small bookplates were much sought after by Sydney society. *Bruce Goold*

Left
Adrian Feint (1894–1971)
Bookplate c1927
12 x 10cm
PRIVATE COLLECTION

Page 25
Grace Cossington Smith
(1892–1984)
Waratah c1930
Oil on board 16.6 x 17cm
ALAN LLOYD COLLECTION

WARATAH SUITE

To quote Bruce Goold (1948–), an Australian artist who specialises in linocuts and has featured a waratah in some form every year since first exhibiting in 1979:

> Anyone who might produce a linocut or woodcut of a waratah printed black and hand-coloured red and green, would inevitably find the reaction: 'Looks like a Margaret Preston.' She championed the waratah and it became synonymous with her name.

In 1992, to exorcise himself from the 'waratah woodblock' style, Bruce commenced the 'Waratah Suite', with the aim of rendering the waratah in different period styles or art movements:

Naturalistic: embodies the 'woodcut style' revealing to the uninitiated the waratah's appearance in the wild and acting as a beginning for the stylistic journey.

Colonial: represents the earliest engravings. Pressed specimens taken to London were accompanied by crude drawings. Botanical engravers, not having seen the living plant, used artistic license to fill in areas of uncertainty, often creating a fantastic plant with little resemblance to the original.

Symbolism: this could be called 'flaming waratah', symbolising the waratah's ability to 'reincarnate' after a bushfire, a phoenix rising from the scorched earth. It also resembles the ruby glass lamps seen at Australian Shrines of Remembrance. Symbolism was chosen by The Royal Botanic Gardens Sydney as the logo of the exhibition 'State of the Waratah'.

Expressionist: one of the thrills of viewing expressionist woodcuts is the animated, sometimes violent, movement created by the gouge as it carves through the block, isolating imagery and leaving furrows of energy in its wake.

Surrealism: the waratah resembles a living heart. Here it's a quilted cushion resting on a star-shaped carousel, the leaves sprout beaks and feathers gradually forming into birds.

Bruce Goold (1948–)
Waratah Suite 1991:
From far left
Naturalistic
Colonial
Symbolism
Expressionist
Surrealism
Linocuts 60 x 35cm
JENNY-JANE CARPENTER COLLECTION

THE WARATAH IN FOCUS

Max Dupain (1910–1992) was arguably the most important Australian photographer of the 20th century. He loved to photograph Australia: people; places; buildings; and in particular Sydney, where he lived throughout his life. He explained the criticism that his photographs lacked angst by saying:

> Moments of agony are out. Looking back on so much work I have to admit there is a certain serenity, a kind of emotional intensity, which probably reflects the rather wonderful life I have enjoyed as an Australian, here in Australia.

In the 1980s he produced a series 'Flowers by Night' which included 'Waratah and Moon', a favourite with artists and botanists alike and seen by the astute observer to be a clever amalgam of two negatives: the whole seductively adding up to more than the two separate parts.

Charles Kerry (1858–1928) has been described as suave, debonair, an excellent shot, a fine angler, a good horseman and a leading skier. He was also one of the best photographers of his time.

His early adoption of the dry plate process (invented in 1878) made outdoor photography much easier. In 1895 he was commissioned by the New South Wales Government to photograph Corroborees and portraits of Aborigines. He developed a 'squatters service' travelling across Australia by train and packhorse to homesteads all over the State, sending the print orders back from Sydney, and incidentally leaving us a document of life in the bush at that time. In 1903 he went on to publish some of the first picture postcards.

Left
Charles Kerry (1858–1928)
Print from glass plate negative
c1900
Tyrrell collection
POWERHOUSE MUSEUM, SYDNEY

Page 29
Max Dupain (1910–1992)
Waratah and Moon 1982
PRINT: JILL WHITE

WARATAH WARRIOR

Richard Thomas Baker (1854–1941) was nothing if not passionate about the waratah. As a botanist at Sydney's Technological Museum, Baker took a keen interest in the economic potential of Australian native timbers, oils and dried specimens. He published books to encourage their commercial use and promoted them on the world stage at international exhibitions. He was also mad about art, especially in its applied form, and fervently believed that the time was right to appoint the waratah as Australia's national flower.

Baker had, in the course of research for his book *The Building and Ornamental Stones of Australia*, become aware of the predominance of the waratah as an ornamental motif on architecture and all manner of the applied arts. The more he looked, the more he found. The more he talked about it, the more the arts and industry responded. In no time, a waratah movement had been born.

In 1915, Baker published a book *Australian Flora in Applied Art. Part 1: The Waratah*. In it he identified hundreds of items, many that the Technological Museum had acquired for their 'Australian Flora Applied to Art' display, to show the great diversity and potential of the waratah as a decorative motif. From buildings to bookbinding, lace to lamps, wallpaper to windows, waratahs bloomed.

In his travels throughout Australia, promoting the waratah movement, particularly to students, Baker saw a wealth of beautiful work which he was inspired to buy. Many pieces were purchased from art and craft exhibitions; others were commissioned from leading firms.

As debate raged over the choice of a national floral emblem, Baker, known as The Commander in Chief of the Waratah Forces, assembled his troops and fought a spirited campaign to have the waratah declared our national flower. He was eloquent in defeat when the wattle won the day, noting in *The Sydney Morning Herald* on 24 September 1910 that:

> As an individual, I think Australia has let a grand opportunity pass of acquiring an heraldic flower of unsurpassed merit for this special purpose alone — a flower that in its generic name, seen from afar, expresses the hope of all Australians that the glories of this continent shall be a light to the rest of the world.

Margaret Betteridge

Page 30
Collage from top left to right: wood panel, curtain frieze, pokerwork box, tile frieze, ceiling rose, fabric, engraved glass, vase, jardiniere, doily, vase, cabinet

APPLYING THE WARATAH

The first two decades of the 20th century saw 'waratah mania' in full flight. True devotees of the waratah who weren't content to enjoy them in the wild, could live among their woven, painted, carved and moulded images in their homes. Really serious telopeaphiles could carpet, glaze, plaster, tile, stencil, wallpaper, curtain, light and fence their homes with waratahs.

Many waratah designs translated well into the prevailing Art Nouveau vocabulary and this distinctly Australian iconography helped to create the Federation style. The Wunderlich Patent Ceiling and Roofing Company, whose pressed metal ceiling could withstand even the highest note of Sydney Town Hall's grand organ, employed an English designer Samuel Rowe to create swirling Art Nouveau-styled waratahs in sheets of pressed metal.

Tile manufacturers in England produced decorative slip trailed waratah tiles for fireplace surrounds and stair risers. Dados and painted wall tile panels were also produced, including a large installation for the foyer of Macdonnell House in Pitt Street, Sydney (now demolished) built circa 1912 for Labor Papers Ltd. Closer to home, Mashmans salt glazed tiles were also in demand.

Plaster gardens adorned the ceilings of modest homes and mansions, with waratahs entwined around native and exotic flowers, or stylised as ventilator covers. Home decorators had their choice of many waratah designs for their floors and walls. They could even fence themselves in with ironwork waratahs.

Eventually the novelty of the nationalistic sentiment began to wear off and by the late 1920s the waratah had become an architectural fashion victim.
Margaret Betteridge

From top left
Tile frieze c1912
T.R. Boote, Waterloo pottery
Staffordshire
Rescued from Macdonnell
House, Pitt Street Sydney 1982
BILL BLINCO COLLECTION

Pressed tin ceiling tile c1910

Page 33
Plaster ceiling rose 1895
'Kirnbank' Arncliffe
New South Wales
(demolished 1980)
LYNDHURST CONSERVATION
RESOURCE CENTRE, HISTORIC HOUSES
TRUST OF NEW SOUTH WALES

EXCEEDINGLY NOVEL

There is no finer tribute to the waratah in Sydney than in its decorative application in Sydney Town Hall. Carved in stone, etched in glass, and wrought in metal, the waratah reigns supreme in Sydney's foremost public building, amid a profusion of neoclassical motifs.

Built in the grand tradition of 19th-century civic halls, Sydney Town Hall was possibly the first building to feature the waratah as an architectural motif. Between 1869 and 1875 the waratah was carved in sandstone on the eastern facade of the building, flanking the civic coat of arms.

Tender documents prepared in 1876 for the vestibule stated: 'The whole of this (ornamental) art work is to be characteristic and the Australian flora and fauna are to be introduced wherever the City Architect may require.' This is believed to be the first reference to the specific use of Australian floral motifs in architectural decoration. Waratahs can be found among the ferns and cabbage palms, the lyrebirds and cockatoos on ten panels of embossed (etched) glass, described by the *Illustrated Sydney News*, July 13, 1878, as 'exceedingly novel'.

Waratah fever was to reach new heights during the construction of the second stage of the Town Hall, when the Main Hall was added to celebrate the Centenary of the founding of the colony of New South Wales. The outpouring of a 'coming of age' sentiment was no doubt the inspiration for the architectural decoration, including the stair railings, the light fittings and the spectacular stained glass windows for which the waratah was the predominant floral motif.

The ultimate celebration of the Centenary is to be found in the stairwells flanking the Main Hall and dominated by the stained glass windows designed by Lucien Henry and made by Goodlet and Smith. Here in a dazzling display of colonial sentimentality, the allegorical figure of Oceania, ruler of the southern seas, is surrounded by totemic waratahs, a powerful symbol of supremacy. *Margaret Betteridge*

From top left
Wrought iron metal railing
c1889
Sydney Town Hall

Light fitting c1889
Sydney Town Hall

Page 35
Lucien Henry
Australia window c1889
Sydney Town Hall

1788 NEW SOUTH WALES 1888
ADVANCE AUSTRALIA
1788-1888
1788-1888
ADVANCE AUSTRALIA

WARATAH FEVER

The demand for waratah wares exploded in the post-Federation glow of Australian nationhood. With few local industries capable of meeting this demand, ceramic firms like Doulton and Co. and the Worcester Royal Porcelain Company saw new business opportunities in their Australian markets.

However, the waratah was no ordinary flower, and the dried specimens, scientific accounts and detailed botanical drawings that were sent back to England were of little help to ceramic artists and modelers. Some created waratahs that looked more like overblown roses or peonies. Others never quite realised the intensity of its natural colours, while still others had difficulty in dealing with its symmetry and scale.

Doulton and Co. went to extensive lengths to get it right, enlisting the help of their Sydney agent, John Shorter, in collaboration with R.T. Baker. Whether it was Louis Bilton's dreamy hand coloured transfer printed versions, inspired by the memory of his visit here to sketch for *The Picturesque Atlas of Australia*, or colourful faience or flambé interpretations, copied from botanical paintings, Doulton's waratahs had class. So too did Moorcroft's richly coloured slip trailed waratahs, and Worcester's waratahs, framed within soft grey acid etched borders.

Shorter's daughter, Lucie (Lulu), didn't let the argument that the waratah was too large a flower to adapt to something so small and delicate as a bone china teacup and saucer phase her, proving with her self titled 'Lulu' series that scale was never a problem.

Home grown china painters decorated imported blanks with waratahs not just in support of the national style idea, but because the waratah adapted so well to the Art Nouveau style of the time. Although china painting appealed particularly to women members of arts and crafts societies, one of its most elegant exponents in Australia was Lawrence Howie who taught at the Adelaide School of Art. *Margaret Betteridge*

From top right
Jardiniere c1900
Doulton & Co. Staffordshire
England
Earthenware
POWERHOUSE MUSEUM, SYDNEY

Porcelain vase c1910
Painted by Lawrence J. Howie
Adelaide
POWERHOUSE MUSEUM, SYDNEY

Exhibition vase 1939
Moorcroft England
PRIVATE COLLECTION

Page 37 from top
Jardiniere, NSW native flora decoration c1892–1902
Unknown maker
(Louis Bilton decorator)
Porcelain
44 x 37cm diameter
ART GALLERY OF NEW SOUTH WALES
PHOTO: RAY WOODBURY FOR AGNSW

Tea service, the Lulu pattern with waratah border c1912
Doulton & Co., established
Staffordshire, England 1854
Bone china transfer printed
and painted various sizes
NATIONAL GALLERY OF AUSTRALIA,
CANBERRA

The use of the waratah as a decorative motif on glass has been limited by the techniques available to the craftspeople. Few examples survive, due in part to the fragile nature of glass and the lack of a flourishing local glass industry until the 1920s, but also because of the difficulty in successfully translating the natural form to the media. For a flower of such intense colour and dimensionality, waratahs on glass were to prove the greatest challenge.

The earliest glass waratahs were engraved on imported glass blanks, inspired by the work of Frank Piggott Webb (b.1859). A master engraver with Thomas Webb, he came to Australia to demonstrate the art form at the 1879 Sydney International Exhibition, eventually settling permanently in Sydney.

Although Lucien Henry proposed a fanciful drawing of a 'versatile conception', a glass decanter in the shape of an upturned waratah, his idea was never translated into a reality. However, the Crown Crystal Glass Company recognised the potential of waratahs to sell their wares, introducing a popular moulded glass 'waratah' pattern in a range of jugs, tumblers, comports and butter dishes in the 1920s and 30s, and 'Waratah' cut glass in 1935.

Crown Crystal used the waratah for moulded decoration on its carnival glass from 1926, the brilliant chemical iridescence strangely at odds with the natural beauty of the flower.

Stained and leadlight glass offered the greatest potential to devotees of the waratah for here was the opportunity to portray the vibrant red flowers and the rich green leaves, as demonstrated in the windows of Sydney Town Hall and in Federation houses throughout Sydney. *Margaret Betteridge*

From left
Waratah pattern jug c1925
Moulded glass
Crown Crystal Glass
Company

Carnival glass plate c1925
Crown Crystal Glass
Company

Page 39
Engraved glass tumbler
c1910
Possibly Frank Piggot Webb

THE RIGHT FLOWER AT THE RIGHT TIME

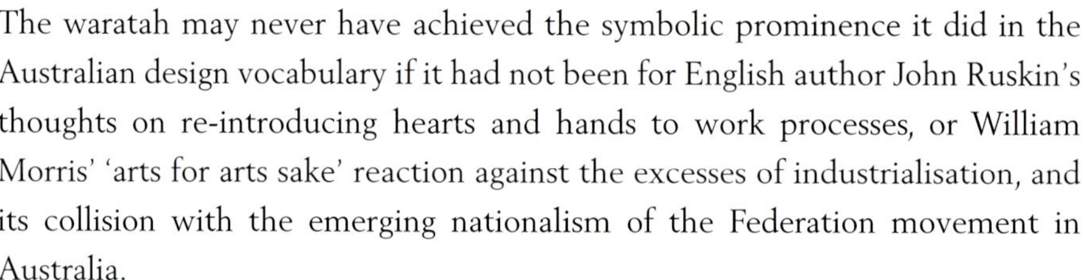

The waratah may never have achieved the symbolic prominence it did in the Australian design vocabulary if it had not been for English author John Ruskin's thoughts on re-introducing hearts and hands to work processes, or William Morris' 'arts for arts sake' reaction against the excesses of industrialisation, and its collision with the emerging nationalism of the Federation movement in Australia.

Australian art and technical schools opened their doors in the late 19th century to men and women eager to learn a craft or trade. The traditional classes in painting and drawing were offered, but so too were wood carving, metalwork, enamelling, leatherwork, weaving, china painting, pottery, jewellery and embroidery.

Art and craft societies were established across Australia to nurture and promote this talent, their members, many of whom were women, encouraged to look no further than the Australian flora and fauna for their inspiration.

Eirene Mort was a champion of the waratah, using it as the logo for the Society of Arts and Crafts of New South Wales, of which she was a member, and as a motif for her many designs. Marian Munday's pottery was the first to be purchased by an Australian gallery and she, like many of her contemporaries, established her own workshop and retail outlet.

Although waratah wood is seldom found in Australian cabinetwork, the flower and leaves of the waratah were popular decorative motifs for woodwork because of their strong definition and their ability to adapt to the prevailing Art Nouveau style. Pyrographers seemed to have the most fun with waratahs. Using black and red Indian ink they coloured their burnished designs to dramatic effect on trays, vases, doily covers and souvenirs. *Margaret Betteridge*

From top right
Society of Arts and Crafts souvenir books cover design by Eirene Mort 1931 & 1956
COLLECTION: SOCIETY OF ARTS AND CRAFTS OF NEW SOUTH WALES

Marian Munday (1850–1935)
Vase 1911, 27.5 x 9.5cm
KEITH FREE COLLECTION

Victorian tea cosy c1890
Embroidered silk
COLLECTION OF THE EMBROIDERERS GUILD OF NEW SOUTH WALES

Coal miners lunch box c1940
Carved silky oak
KEITH FREE COLLECTION

Pokerwork box c1940
PRIVATE COLLECTION

Page 41
Arts and crafts chest c1925
Handpainted 105 x 50 x 54cm
PRIVATE COLLECTION

A SISTERHOOD OF QUEENS

Of all the Australian flora, the waratah — the 'glory of the Australian bush', was the most commonly used floral motif. Perhaps it had something to do with its status among the Australian wildflowers, recognised as early as 1840 by artist Louisa Ann Meredith who marvelled at a group in flower as a 'sisterhood of queens'. Nuri Mass writing in *Australian Wildflower Magic* claimed 'There's none so proud and stately and grand as Waratah.'

In 1907, the city of Melbourne hosted the Australian Exhibition of Women's Work. It attracted thousands of entries from the world over but displayed a predominance of work by Australians. Eirene Mort designed a twelve foot square heraldic carpet featuring the waratah, which was worked by members of the New South Wales Exhibition Group and presented to Lady Northcote, the wife of the Australian Governor General and patron of the exhibition.

The sisterhood continued to stitch and sew, creating not castles, but a little pin money, and homes filled with embroidery, crochet, lace and other fancywork. Doilies were by far the most popular product of their labours and designs were limited only by the imagination and skill of the creator. Some of the highly talented members of the arts and crafts societies, such as Muriel Cornish, produced needlework of an exceptionally high standard and examples of her work illustrate the design opportunities afforded by the waratah.

For many years it seemed that the best inspiration the waratah could hope to be in textile and fabric craft was as a pattern for domestic needlework on the pages of women's magazines. Francis Burke (1907–1994), however, gave the waratah another identity as a motif for printed fabric. Her company, New Design Pty Ltd, heralded a rebirth for the waratah when she featured it in her range of modernist fabrics, used by leading architects of the day.

Margaret Betteridge

From top right
Muriel Cornish
Doily with waratah design
1919
Embroidered linen
58 x 81.5cm
ART GALLERY OF NEW SOUTH WALES

Francis Burke Fabrics
Established Australia 1937
Waratah length of fabric
c1955
Screenprint on cotton
83 x 120cm
NATIONAL GALLERY OF AUSTRALIA, CANBERRA

Madeleine King
Curtain frieze with waratah, thistle, clover and rose motif
1910
Hand-stenciled linen
135.7 x 39.4cm
ART GALLERY OF NEW SOUTH WALES
PHOTO: RAY WOODBURY FOR AGNSW

Page 43
Artist unknown
Piece of textile waratah pattern c1890
Cotton 74.4 x 37cm
NATIONAL GALLERY OF AUSTRALIA, CANBERRA GIFT OF KEITH FREE, 1984

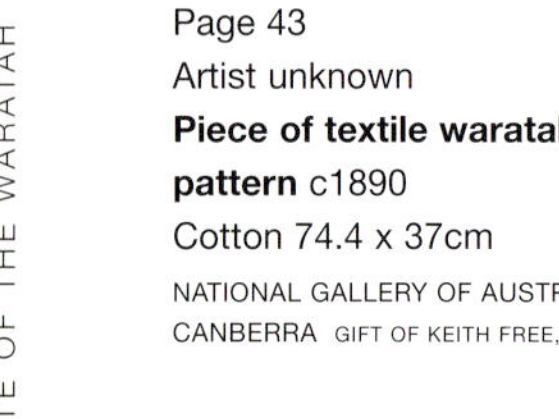

WEARING THE WARATAH

'The Blue Mountains are where my dreaming comes alive — the waratah is my totem in nature: red, passionate, rare.' Artist Jenny Kee returned from 'Swinging London' in 1972 and took a fresh look at Australia. She began to design knits and textiles which incorporated distinctly Australian motifs and created a new fashion style. The mixture of rich ethnic embroideries and retro clothing she had been working with overseas became a trigger for her own design aesthetic. Knit and fabric were a canvas for her artworks. Each design began as a series of small paintings which she collaged into larger stories or themes.

'Flamingo Park', her shop in Sydney's famous heritage arcade, The Strand, became the place to buy and to be seen. The 'international set', and an Australian public hungry for a fashion style to call their own, flocked there for 20 years. The Flamingo Park fashion parades, where waratahs decorated the stage as well as the clothes, were the highlight of the Sydney season, involving like-minded artists and designers, and delighting her public.

Today Jenny Kee lives in the Blue Mountains on the edge of a 'prehistoric forest' with her beloved flowers growing all around her. Here she continues to be one of our most enthusiastic promoters of the waratah, celebrating it in her art, design and writing.

From left
Jenny Kee (1947–)
**Woman's beach ensemble,
Waratah and Blackboy**
1988–1989
Printed silk and lycra
POWERHOUSE MUSEUM, SYDNEY
PHOTO: MONTY COLES FOR VOGUE AUSTRALIA

**Jenny Kee with waratah,
Blue Mountains** 1999

Page 45
Jenny Kee
Waratah and Blackboy scarf
1984
Printed silk
132 x 136cm
COLLECTION OF ARTIST

THE WARATAH FROCK

The individual and majestic beauty of the native waratah inspired me to create a beautiful gown that became a feature of every collection. The shapes of the flower petals and leaves are cut and stitched; the colours and shapes are painted, printed and embroidered on different textile surfaces; the fabrics are torn, cut, swathed, draped and twisted — to reveal the essence of the waratah.

Linda Jackson sees her clothes as an integration of art, life and environment. In the early 1970s she travelled widely: Papua New Guinea, Asia and Europe. The resultant textiles and shapes were inspired not only by the beauty and colour of our unique Australian environment but also by tribal cultures and traditional haute couture.

Linda has designed a more comprehensive range of hand-printed waratah fabrics than any other Australian artist. She collaborated with Jenny Kee in Flamingo Park in the 1970s and in the early 80s opened her 'Bush Couture' Studio, showcasing her flamboyant hand-painted textiles, fabulous frocks and opal jewellery.

Linda's passion for the waratah has been expressed in a myriad of ways. Her wanderlust has been satisfied by living in the bush in remote Australia and working and teaching in Aboriginal communities. Now, the rainforests of Far North Queensland have captured her imagination in a new range of furnishings, wallpapers and prints and the Torch Ginger has become her new waratah.

Left
Linda Jackson (1950–)
Waratah scarf 1990
Screenprinted silk
88 x 88cm
Commissioned by Oroton
COLLECTION OF ARTIST
PHOTO: FRAN MOORE

Page 47
Linda Jackson
Blue Mountains Waratah Frock
1983
Silk organza, taffeta and chiffon
COLLECTION NATIONAL GALLERY OF VICTORIA
PHOTO: FRAN MOORE

CELEBRATIONS AND INVITATIONS

The opening of Sydney's Centennial Hall was, of course, supposed to take place in 1888 as a celebration of the centenary of the colonisation of New South Wales. However, construction fell behind and the opening took place a year later. It was a grand occasion with spectators filling every inch of space in the galleries and hall for the midday launch, and spectacular ball in the evening for 2800 guests. The invitation was splendid, featuring swirls of waratahs and decorated with gold leaf. The style and the visual references to the stained-glass windows in the Town Hall lead us to assume that it was designed by Lucien Henry who had recently completed work on the new building and was keen to promote the use of the waratah in the decorative arts.

A few years later, the pyrotechnic illumination of Sydney Harbour, as part of the Inaugural Celebrations at Sydney, must have been a spectacular contribution to the celebration of the birth of a Nation. On 1 January 1901, a quarter of a million people strolled through the brilliantly decorated streets of Sydney and, in the afternoon, flocked to Centennial Park to see the formal proclamation by the Governor General, Lord Hopetoun, reading from the Queen's text. The ordinary people of Australia were seeing the fulfilment of a Federal dream: 'One people, one flag, one destiny'. They would also have enjoyed the illuminations which might be seen by all, whether or not they had an invitation.

Federation and all the events surrounding it in Sydney and Melbourne produced a dazzling array of invitations, menus and programs. They are revealing in their use of colonial and national icons and symbols, including the new coat of arms. Typically, a rose, signifying our close ties with England, was included and, pre Harbour Bridge and Opera House, the panorama of North and South Heads offered the most spectacular view of the glory of Sydney Harbour. The waratah, not yet having lost to the wattle as our national flower, featured above all others.

Page 49 from top
Invitations for:

Ball at Sydney Town Hall 1889

**Pyrotechnic Celebrations
Sydney** 1901

NATIONAL LIBRARY OF AUSTRALIA,
CANBERRA

THE WARATAH AT WAR

> May the Waratahs grow in strength, flourish whilst temporarily transplanted in foreign soil, and return to us rich in achievement for King and country no less conspicuous and brilliant than their namesake, the crimson monarch of the Australian bush. PALMER ADDRESS, 1915

A crowd of 2000 farewelled the 50 'Waratahs' as they left on a recruiting march from Nowra on 30 November 1915, and applauded the address by Mr Palmer, organising secretary of the State Recruiting Committee. Their number had swelled to 110 when they arrived in Sydney in pouring rain on 16 December. Whilst the numbers were insignificant in the context of Australia's contribution to the war, recruiting marches were morale boosting and involved the whole community in a patriotic spectacle.

In the lead up to the march, the local Red Cross branch offered to contribute a suitable banner described in the local paper:

> In the centre of the banner, a Waratah with leaves is nicely worked in red and green silk. The work was undertaken by Mrs Dr Rodway and Miss McIlwraith who are to be congratulated upon their artistic designing, and taste displayed in work.

After the presentation of the banner and the message from Mr Palmer. the band played several patriotic pieces, and after cheers for The Waratahs, the program drew to a close with the singing of the national anthem. *Alan Clark*

From top
Embroidered cloth badge
PRIVATE COLLECTION

Postcard celebrating the arrival of the American Fleet in Sydney 1908
NATIONAL LIBRARY OF AUSTRALIA, CANBERRA

Page 51
Photo **The Waratahs cross the Nowra Bridge on the first leg of their march** 1915
SHOALHAVEN HISTORICAL SOCIETY

SOUTH COAST WARATAHS

WISH YOU WERE HERE

The postal card system introduced to Australia in 1875 was a great success and by the early 1900s picture postcards were at the height of their popularity. Postcards were useful for letting your friends know that you were 'on vacation' and were proudly collected in albums. This new art form attracted a wide range of designs. Photographer Charles Kerry translated his most popular images into postcards from 1903 onwards and botanical artist Margaret Flockton's 'Art Series' cards published in 1908 were sold at 1 shilling & 5 pence for a set of 12.

In the 1930s a cruise on an Orient Royal Mail steamer was the most fashionable vacation. Sir Colin Anderson, director of The Orient Line from the 1920s–50s was also a director of London's Tate Gallery and commissioned designs from Australian painters, including Douglas Annand and later Donald Friend, who he found to be 'fresh', and 'not hemmed in'. The menu covers became treasured souvenirs.

After the war, travel was all the rage again. The Blue Mountains, west of Sydney, were stylish, within easy reach, and famous for their scenic walks and waratah viewing. The availability of the screen printing process, only newly arrived in Australia, led to a range of attractive inviting posters. Australia was on the map.

From top right
Australian postage stamp 3/–
15 July 1959
Design: Margaret Stones

Australian postage stamp 30c
10 July 1968
Design: N. Wilson

Orient Line menu c1938
NATIONAL LIBRARY OF AUSTRALIA, CANBERRA

New South Wales postcard
Issued 1887–1888

Group of picture postcards
Photos by Keys c1910
BILL BLINCO COLLECTION

Page 53
Henry Rousel (b.1897)
See for yourself ... The Blue Mountains c1955
Silkscreen 102 x 76cm
JOSEF LEBOVIC GALLERY

See for Yourself...
The
BLUE MOUNTAINS
OF NEW SOUTH WALES
For Information .... WRITE TOWN CLERK KATOOMBA N.S.W. AUSTRÁLIA
H. ROUSEL

WARATAHS THAT WORK

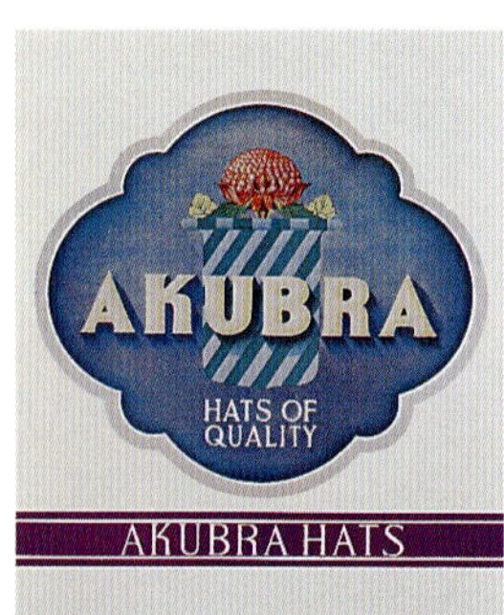

The waratah was officially designated the State flower in 1962 but long before this, clubs and associations realised the power of the waratah to symbolise. 'The State of New South Wales'. The strip of the New South Wales' Rugby team, 'The Waratahs', not surprisingly has carried the floral emblem since their first official gallop in the 1880s, and it has enhanced the mason's regalia since the New South Wales Lodge was formed in 1888.

State government departments, major stores and small companies have used the waratah in their advertising and promotions. Shelleys, established in Broken Hill in 1884, have always featured it on their soft drink bottles and Grace Bros store chose a stylised version of the waratah when they revamped their company image in the 1980s.

Matchboxes, flour bags and paperweights have all used waratahs as advertising and decoration but tins win the race 'by a country mile'. Griffiths' tea tins with their Art Nouveau decorations, were favourites for kitchen storage. Arnotts, when not using their signature parrot logo, have featured the waratah on biscuit and cake tins since they first manufactured fancy tins in the early 1900s.

Once an Akubra hat was sold in a box resplendent with a waratah and a waratah in an advertisement assured us that we were 'helping Australia along its path of national destiny'. In 1917 Waratah Motor Spirit was advertised as 'The Spirit that Moves'. More recently, a waratah imprinted on a drivers licence is the New South Wales' motorists lot.

Today, guests of the Premier are reminded they are in New South Wales when they find themselves walking on a striking waratah emblazoned carpet.

From top right
Brass paperweight
Ammonia Company of
Australia, Sydney
ROBERT HUTCHINSON COLLECTION

Enamel badge c1940

Hat box label c1930
Akubra collection

**Shelleys lemonade
advertisement** c1930

Wills cigarette card c1920
ROBERT HUTCHINSON COLLECTION

Toffee tin c1940
**Arnott's 3lb Christmas cake
tin** 1957

Page 55
Griffith Tea tin c1910
ROBERT HUTCHINSON COLLECTION

CONTEMPORARY WARATAH

From top
Photo album with painted velvet cover c1940

Katrina Daly (1955–)
Painted wood cut-out, one of pair c1985

Katrina Daly
Cushions 1990

Page 57 top
Vase possibly Pates Pottery c1950
JOHN WADE COLLECTION

Page 57 bottom left
Diana vase c1960
ROBERT HUTCHINSON COLLECTION

Page 57 bottom right
Vase Studio Anna c1950
BILL BLINCO COLLECTION

The Australian pottery industry blossomed briefly from the 1940s onwards. The most popular commercial brand was 'Diana' made in suburban Marrickville from the 1940s–1970s and best known for 'ovenware' and attractive vases. Studio Anna, founded by Karel Jungvirt in the early 1950s, specialised in hand-painted and printed souvenirs, now much in demand by collectors.

From the 1970s, Australian flora on fabric had a revival. Artist Deborah Leser produced her popular 'waradot' furnishings and fashion fabrics using a sophisticated combination of hand painting and resist dying, whilst Helga Muschinsky, Linda Jackson and Jenny Kee each designed the definitive version of the waratah scarf. In 1989, gloriously embroidered waratahs adorned the suit designed by Lee and Stephen Galloway and worn by curator Jane De Teliga to the launch of 'Australian Fashion: the Contemporary Art' in London, Tokyo and Sydney. Australian wildflowers were a favourite with jewellery designer Robyn Gordon; Deborah Leser created waratah gloves; and milliner Peter Jago conceived the ultimate waratah titfer for the Melbourne Cup.

Katrina Daly one of a group of innovative young designers working with silk screen printers 'House of Feral', made fabrics for soft furnishings and printed clothing for 'Salon One'. In the tradition of the arts and crafts movement, Katrina also worked with wood and in the 1980s made an attractive pair of trompe l'oeil waratah window decorations.

The new generation of designers have invented original ways of utilising the waratah. The traditional souvenir lives on, but increasingly its use is in the more sophisticated arenas of architecture and graphic design.

EVEN IN ARCHITECTURE

The waratah is particularly suited to architectural and ornamental use with its brilliant colour and distinctive shape, easily recognised even in highly stylised versions. In architecture, plant forms have often been appropriated into structural elements. The Egyptians adapted the lotus and the Romans the acanthus leaf for column capitals. Gothic cathedrals often incorporated stained-glass 'rose' windows. In the early part of the 20th century, Art Nouveau expressed itself through vine-like tendrils of wrought iron, and Art Deco elaborations sometimes included abstractions of foliage.

The earliest examples of the waratah in architecture on view in Sydney include: the Town Hall (c1870); the Woolloomooloo and Palace Garden Gates (1873 and 1889), both in the Royal Botanic Gardens; The Lands Department Building in Bridge Street (c1890); the newly-restored Post Office in Martin Place (c1890) and the former City Mutual Life Building in Hunter Street (1936).

The early period of the modern movement rejected ornament as being sentimental and degenerate, so its use diminished. Fashions cycle, so in our current phase of mature modernism, ornament may be introduced once again without guilt. A dramatically graphic waratah in terrazzo graced Sydney's Darling Harbour until recently, and Skygarden and Chifley Tower, both in the Sydney CBD, are examples of Australian flora in contemporary commercial design.

In 1998 architect Eric Kuhne, with the Royal Botanic Gardens and landscape design consultants Site Image, developed a garden the size of a playing field at Darling Park, based on the pattern of the waratah. This abstraction, not noticeable at ground level, is nevertheless clearly perceived from above by workers in the adjacent high-rise offices. *Neville Quarry*

From top right
Terrazzo floor insert
Harbourside, Darling Harbour
1988
Design Bruce Goold for Public Art Squad, (demolished 1998)

Palace Garden Gates
Royal Botanic Gardens Sydney
Sandstone pillar 1889 (detail)

Skygarden terrazzo and mosaic waratah
Designed and produced by Public Art Squad 1990

Page 59
Waratah Garden, Darling Park
1998
Developed by Lend Lease

WARATAH RELATIONS

Waratahs are not only spectacular to look at, but have something spectacular to tell us about our land and its flora.

Waratahs belong to a group of closely related plants, descended from an ancestor that lived over 70 million years ago. We know that this ancestor existed then because it left recognisable fossils. Our planet has changed dramatically in many ways since that time. The continents have changed their positions and the climate has become cooler and drier. Back then, Australia was still joined to Antarctica and this 'super-continent' was almost touching South America. There were no polar ice caps and Antarctica was at least partly covered in moist forests. Forests formed an almost continuous habitat linking Australia with South America. The 'ancestral waratahs' grew in these forests and became widespread across the super-continent, the remains of Gondwana.

The rifting of Australia from Antarctica about 60 million years ago and its subsequent drift northwards had a profound impact on the plants and animals of the southern end of the world. The ancestral waratahs of South America and Antarctica were separated from those in Australia, and the two groups evolved in isolation. Later environmental changes caused further differentiation in these lineages, giving rise to the twelve waratahs and 'waratah relatives' that survive today in Australasia and South America.

The waratah's closest relatives — those which can be thought of as its 'sisters' — are the four other species of *Telopea*. They grow in south-eastern Australia from Tasmania to north-eastern New South Wales. Its 'cousins' are four species of *Alloxylon*, or 'tree waratahs', from eastern Australia and Papua New Guinea, two species of *Oreocallis* from Peru and Ecuador, and one species of *Embothrium* from Chile and Argentina. *Peter Weston*

Page 61
Waratah relatives
from top left: to right
Embothrium coccineum
Oreocallis grandiflora
Oriocallis mucronata
Alloxylon pinnatum
Alloxylon wickhamii
Alloxylon flammeum
Alloxylon brachycarpum
Telopea speciosissima
Telopea aspera
Telopea truncata
Telopea oreades
Telopea mongaensis
PHOTOS: E.H. NORRIS, PETER WESTON,
ELIZABETH BROWN, JAIME PLAZA, MIKE CRISP

THE NEW BREED OF WARATAHS

By the beginning of the 20th century, bush burning, indiscriminate picking and encroaching civilisation were threatening the survival of waratahs in the wild. In the 1920s conservationists Percy and Olive Parry, who were passionate about Australian wildflowers, bought land 50 kilometres north of Sydney which later became 'Floralands'.

Here they developed the concept of 'preservation by cultivation', leading the way for the commercial growers who today supply cut flowers to an eager market, and home gardeners who can cultivate the new 'easier to grow' waratahs.

To increase the vigor of the traditional New South Wales waratah, *Telopea speciosissima*, it has been crossed with other species resulting in 'Shady Lady' and 'Corroboree', for instance, which are both very hardy and have profuse flowers. Plants are also selected for their large flowers and longer vase life, such as 'Fire and Brimstone' and 'Songlines', or for the pretty pink tips on the flowers as with 'Brimstone Blush'. 'Wirrimbirra White' lacks the gene necessary for a red flower and was named after the sanctuary near Bargo, New South Wales where it was first cultivated in the late 1970s. 'Olympic Flame', developed by Sydney University, is a splendid bright-red specimen with a sweet honey nectar described as 'rocket fuel for native birds'.

A charity auction held at the Sydney flower markets each spring celebrates 'White Waratah Day' and signals the start of the waratah season which peaks in early October. Now we can enjoy waratahs in our homes and gardens whilst ensuring that future generations can see the breathtaking sight of waratahs in the wild. *Cathy Offord*

From top
'Songlines'
'Wirrimbirra White'
'Fire and Brimstone'
'Brimstone Blush'
PHOTOS: JAIME PLAZA AND YELLOW ROCK NURSERY

Page 63
Telopea speciosissima 1999
Photo Gary Heery

ACKNOWLEDGEMENTS

EDITOR ROSIE NICE

PUBLICATION DESIGN ROBERT WIRTH

PHOTOGRAPHS JAIME PLAZA FOR ROYAL BOTANIC GARDENS SYDNEY
UNLESS FROM OTHER INSTITUTIONS OR OTHERWISE CREDITED

SPECIALIST WRITERS

LEO SCHOFIELD, DIRECTOR OLYMPIC ARTS FESTIVAL

RICHARD NEVILLE, CURATOR OF PICTURES RESEARCH, MITCHELL LIBRARY

MARGARET BETTERIDGE, CURATOR SYDNEY TOWN HALL COLLECTION

NEVILLE QUARRY AM, EMERITUS PROFESSOR OF ARCHITECTURE

PETER WESTON, SENIOR RESEARCH SCIENTIST, ROYAL BOTANIC GARDENS SYDNEY

CATHY OFFORD, HORTICULTURAL RESEARCH OFFICER, ROYAL BOTANIC GARDENS SYDNEY

BRUCE GOOLD, AUSTRALIAN ARTIST

ALAN CLARK, HISTORIAN AND AUTHOR *THE WARATAHS*

REFERENCES

JANE DE TELIGA — *AUSTRALIAN FASHION – THE COMTEMPORARY ART*

GLYNIS JONES — *JENNY KEE: COLLECTING THE ARCHIVES OF AN ICON* POWERLINE WINTER 1999

MARGARET BETTERIDGE — *AUSTRALIAN FLORA IN ART*

MARGO BEASLEY — *SYDNEY TOWN HALL, A SOCIAL HISTORY*

DAVID COOK — *PICTURE POSTCARDS IN AUSTRALIA 1898–1920*

DAVID P MILLAR — *CHARLES KERRY'S FEDERATION AUSTRALIA*

EDWIN WILSON — *THE WISHING TREE*

PAUL NIXON — *THE WARATAH*

C.W. PECK — *AUSTRALIAN LEGENDS* (1925)

SASKIA HAVEKES AND GARY HEERY — *GRANDIFLORA*

SYDNEY TOWN HALL — *SPECIFICATION OF WORK REQUIRED IN THE COMPLETION OF THE ANTE CHAMBER CRS 65/1210, 1876.*

THANK YOU

TO GALLERIES AND MUSEUMS WHO GENEROUSLY OPENED THEIR WARATAH ARCHIVES, IN PARTICULAR THE POWERHOUSE MUSEUM SYDNEY, THE HISTORIC HOUSES TRUST OF NEW SOUTH WALES, THE NATIONAL GALLERY OF AUSTRALIA, THE ART GALLERY OF NEW SOUTH WALES, THE STATE LIBRARY OF NEW SOUTH WALES, THE NATIONAL LIBRARY OF AUSTRALIA AND THE AUSTRALIAN WAR MEMORIAL

TO THE EXHIBITION TEAM AT THE NATIONAL TRUST S.H. ERVIN GALLERY
EXHIBITION DESIGN BANNYON WOOD
EXHIBITION GRAPHICS WIRTH VISUAL COMMUNICATIONS

TO THE FRIENDS OF THE GARDENS AND STAFF AT THE ROYAL BOTANIC GARDENS SYDNEY WHO HELPED THIS PROJECT GROW — A SPECIAL THANK YOU

© 2000
ROYAL BOTANIC GARDENS SYDNEY
AND NICE EXHIBITIONS

ISBN 0 7347 2024 6

Cover
James Sowerby (1757–1822)
Embothrium speciosissimum 1793
Hand coloured engraving (detail) 23.5 x 16cm
ROYAL BOTANIC GARDENS SYDNEY

Inside cover
Waratah fabric c1955
Francis Burke Fabrics
established 1937
NATIONAL GALLERY OF AUSTRALIA, CANBERRA